Living Green

Saving Water

By Meg Gaertner

level 2 little blue readers

www.littlebluehousebooks.com

Little Blue House is distributed by North Star Editions:
sales@northstareditions.com | 888-417-0195

Produced for Little Blue House by Red Line Editorial.

Photographs ©: Shutterstock Images, cover, 4, 6–7, 9, 10 (top), 10 (bottom), 13, 14–15, 17, 18–19, 20, 23 (top), 23 (bottom), 24 (top left), 24 (top right), 24 (bottom left), 24 (bottom right)

Library of Congress Control Number: 2022901948

ISBN
978-1-64619-601-2 (hardcover)
978-1-64619-628-9 (paperback)
978-1-64619-679-1 (ebook pdf)
978-1-64619-655-5 (hosted ebook)

Printed in the United States of America
Mankato, MN
082022

About the Author

Meg Gaertner enjoys reading, writing, dancing, and being outside. She lives in Minnesota.

Table of Contents

Saving Water

People drink fresh water. They also use it for cleaning, growing plants, and raising animals.

If people use too much, fresh water can run out. So, people must save water. Saving water means not wasting it.

Water must be cleaned before people drink it. Cleaning water takes money and energy. So, saving water can save money and protect Earth.

faucet

Saving at Home

You can save water in many ways. You can take short showers. You can turn off the faucet when you brush your teeth.

You can save water when doing the laundry. Only use the washer when you have a full load of clothes.

washer
clothes

You can save water when doing the dishes.
Only use the dishwasher when you have a full load of dishes.

dishwasher
dish

You can also fill the sink with hot water and soap. Wash the dishes in the soapy water. Only turn on the faucet to rinse the dishes.

Sometimes faucets drip.
If a faucet starts dripping,
tell an adult.
They can fix the faucet so
it does not waste water.

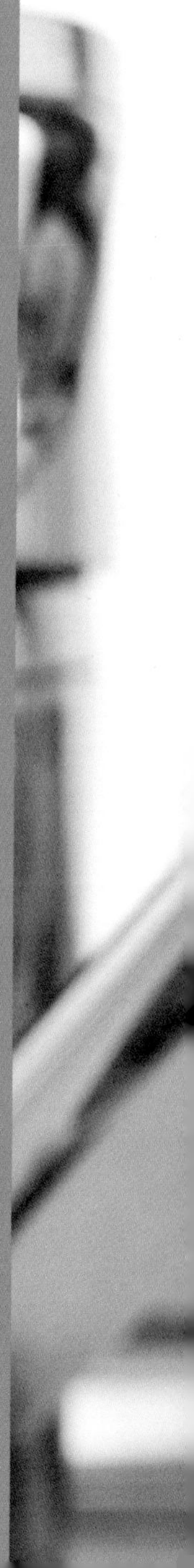

drip

rain barrel

Saving Outside

Plants in the garden need to be watered.

To save water, use a rain barrel instead of a hose.

When it rains, the rain barrel fills with water. Save it for dry days. Use it for garden plants. Everyone can help save water.

Glossary

dripping

rain barrel

faucet

washer

Index